YOUR KNOWLEDGE HAS VALUE

Deepwater Horizon. An Analysis of the Political and Legal Consequences of the Biggest Oil Spill in U.S. History

Stella Benedickt

Bibliographic information published by the German National Library:

The German National Library lists this publication in the National Bibliography; detailed bibliographic data are available on the Internet at http://dnb.dnb.de.

ISBN: 9783346769039
This book is also available as an ebook.

Print and binding: Books on Demand GmbH, Norderstedt, Germany
Printed on acid-free paper from responsible sources.

The present work has been carefully prepared. Nevertheless, authors and publishers do not incur liability for the correctness of information, notes, links and advice as well as any printing errors.

GRIN web shop: https://www.grin.com/document/1303027

Christian-Albrechts-Universität zu Kiel

Deepwater Horizon – An Analysis of the Political and Legal Consequences of the Biggest Oil Spill in U.S. History

Student: Stella Benedickt

Study program: Internationale Beziehungen und Internationales Recht
Seminar: Current Developments in Ocean Governance WS 19/20

Modul: Internationale Beziehungen
Date: 28.09.20

Inhaltsverzeichnis

1. Introduction

For centuries the oceans are a source for food supply, transportation, security, oil and gas resources. At the same time, human pressure the oceans both from land and see and climate change causes further tension. Ocean governance developed as a response to this. It's the conduct of the policy, actions and affairs regarding the oceans and is engaged for example with the division of the ocean space and the diverging regulations to control human activity in the oceans. It strives to resolve governance problems that arise from the zonal and sectoral aspects of the ocean spaces through a set of norms. Over the centuries coastal waters developed as regions under state control, whereas many high oceans remained relatively open-access areas. Costal states gained sovereignty rights and controlled the marine resources.[1] By the middle of the twentieth century there was an impetus to extend national claims over offshore resources. The exploitation fever grow as technology opens new ways to tap the resources of the sea. Especially oil exploration was moving further from land and deeper into the bedrock of continental margins in the late 1960s. Oil drilling equipment in the Gulf of Mexico was already going as far as 4,000 metres below the ocean surface. The fast development of the offshore oil production needs to be seen in a political context as till the 1970s the United States had been the world's number one producer of petroleum and American oil companies determined the global price of oil. However, the American demand for oil steadily increased and so did the overall reliance on imports. The "oil shocks" in 1973 and 1979 were the clearest and most painful affirmation of the nation's vulnerability because of its dependence on import oil.[2] The more oil the U.S. produces itself the more the country can limit its involvement with foreign powers and concentrate on entirely meeting its own energy needs. Offshore drilling and shale oil will provide that quantity of oil and any additional oil production can be used to trade internationally and thereby making the U.S more powerful on the global stage.[3] "[…] a country can make itself better off by restricting the imports of particular product if that country has a degree of market or 'monopsony' power […]."[4] This is especially true for oil as much of the supply comes from government-controlled entities. Dependence on the other hand can threaten U.S. national security and revenues may undermine the efforts to promote good governance.[5] The Gulf of Mexico is one of the most important regions for energy resources and infrastructure for the U.S., both onshore and offshore.

---

[1] Singh, Pradeep A. and Ort, Mara „Law and Policy Dimensions of Ocean Governance", 2020, p. 45 f.
[2] Bacevich, Andrew J. „The limits of power", 2008, p. 31.
[3] https://www.tigergeneral.com/advantages-offshore-oil-rigs-drilling/
[4] Krupnick, Alan; Campell, Sarah; Cohen, Mark A. and Parry, Ian W.H. „Understanding the Costs and Benefits of Deepwater Oil Drilling Regulations", 2011, p. 33.
[5] See Ibid., p. 33 f.

The offshore oil production accounts for 17% of total U.S. crude oil production, over 45% of total U.S. petroleum refining capacity is located along the Gulf coast.[6] The history of the offshore oil and gas industry in that region is marked by environmental, social and political challenges.[7]

There are key events in the history of the oil industry that cause the onset of major changes. One of these events is the explosion at the Deepwater Horizon mobile offshore drilling unit in the Gulf of Mexico on April 20, 2010.[8] This paper shortly explains the Incident and it's environmental and economical consequences. However, the focus is on the political and legal changes that can be seen as a result of the accident. Who was held responsible for the spill? What was the direct response from the Obama Administration and how did the biggest oil spill in history changed policy concerning safety and oil production? In the last chapter Trumps oil policy will be presented together with the perspectives of the offshore industry.

2. Deepwater Horizon

Before showing what happened on April 20, 2010 the author will describe the history of oil drilling in the ocean and the legal framework, regulations, norms and environmental conditions that pour the drilling activity. The focus will be on the Gulf region as this is where the incident happened.

2.1 Historical Background

The Gulf of Mexico is an arm of the Atlantic to the south of the United States and to the east of Mexico. The U.S. states of Texas, Louisiana, Mississippi, Alabama and Florida border the Gulf on the north. The offshore industry developed in that region as the gently sloping outer continental shelf allowed to move slowly into deeper water. It evolved from the onshore industry: after extending the oil research through wetlands and lakes the industry reached the outer continental shelf.[9] 1938 the Creole field, about a mile offshore in 14 feet of water, became the first producing property in the gulf. In the 1920s/ '30s fossil fuels were first extracted from marshes and bayous of southern Louisiana. New drilling techniques and specialized barges to hold the drilling derricks were developed to make the drilling in deeper water possible.[10]

---

[6] https://www.eia.gov/special/gulf_of_mexico/
[7] Bureau of Ocean Energy Management „The Offshore Petroleum Industry in the Gulf of Mexico: A Continuum of Activities" p. 1.
[8] James, Robert A. and Pulman, Stella „Deepwater Horizon: A Decade of Legal Impacts", 2020, p. 1.
[9] Bureau Ocean Energy Management „The Offshore Petroleum Industry in the Gulf of Mexico: A Continuum of Activities" p. 1.
[10] https://fortune.com/2011/01/24/a-short-history-of-drilling-in-the-gulf-of-mexico/

After World War II., the offshore oil production experienced an enormous boom, due to a high demand for oil.[11] "Oil companies had to wait for the technology" Bill Jackson, a Regional Sales Manager is remembering that time.[12] Till today the Gulf of Mexico remains the primary laboratory for technological innovation and regulatory practices in the offshore industry worldwide.[13] In 2009 the first platform drilled oil and gas from the Lower Tertiary and by 2010 the industry had announced 19 oil field discoveries in the Lower Tertiary providing million barrels of oil equivalent. However, drilling in the ultra-deepwater comes with technical challenges as the subsalt geology is unique. Shut-in pressure is high, and the bottom-hole temperature can exceed 350-degree Fahrenheit. Furthermore, salt and tar-zone formations can be problematic and ensuring hydrocarbon flow through riser and pipeline can be difficult. Nevertheless, the challenges seemed manageable and the rewards worth the risk. Government and industry were optimistic about the potential of offshore drilling for the nations future.[14]

Despite, Oil spills caped the euphoria of new achievements and continues profit. 1969 the Santa Barbara oil spill gave birth to the modern environmental movement and changed the trajectory of oil and gas exploration. The blowout was caused by inadequate safety precautions, ca. 3-million gallons of crude oil spewed into the ocean creating an oil slick along California's coast that killed thousands of birds, fish and sea mammals.[15] 1989 another fatal oil spill made history: the Exxon Valdez oil tanker sank in the Price William Sound, a bay of the Gulf of Alaska and spilled ca. 11-million gallons oil. The tanker struck a well-known navigation hazard, the impact of the collision tore open the ship's hull and the oil spilled into the water. Like the Santa Barbara spill, this accident killed thousands of animals, including birds, otters, seals, eagles and whales.[16] The catastrophe lead to the Oil Pollution Act 1990, which intended to avoid oil spills from vessels and facilities, enforce removal of pilled oil and assigned liability to the cost of clean-up and damages.[17] Until the Deepwater Horizon oil spill, the Exxon Valdez was the worst oil spill in the U.S. history.[18]

---

[11] https://fortune.com/2011/01/24/a-short-history-of-drilling-in-the-gulf-of-mexico/
[12] Bureau Ocean Energy Management „The Offshore Petroleum Industry in the Gulf of Mexico: A Continuum of Activities" p. 5.
[13] Bureau Ocean Energy Management „The Offshore Petroleum Industry in the Gulf of Mexico: A Continuum of Activities" p. 1 ff.
[14] National Commission on the BP Deepwater Horizon Oil Spill and Offshore Drilling „Deepwater – The Gulf Oil Disaster and the Future of Offshore Drilling", 2011, p. 51 f.
[15] https://www.latimes.com/local/lanow/la-me-ln-santa-barbara-oil-spill-1969-20150520-htmlstory.html
[16] https://www.history.com/topics/1980s/exxon-valdez-oil-spill
[17] https://www.boem.gov/sites/default/files/documents//The%20Oil%20Pollution%20Act%20of%201990.pdf
[18] https://www.history.com/topics/1980s/exxon-valdez-oil-spill

2.2 Legal Framework and structures

The first codification of international maritime law was 1958. By that time four conventions were concluded, concerning the high seas, the coastal sea, contiguous zone and the continental shelf as well as deep-sea fishing. Nevertheless, there were still questions unanswered and issues excluded, like the breadth of the costal see or the codification of pollution control.[19] 1982 the United Nations Convention on the Law of the Sea (UNCLOS) was signed after decades of negotiations and codified issues such as navigational rights, territorial sea limits and economic jurisdiction.[20] Due to the UNCLOS every country with access to the ocean owns an exclusive economic zone (EEZ) that is an area beyond and adjacent to the territorial sea (Art. 55 Part 5 UNCLOS). The exclusive economic zone shall not extend beyond 200 nautical miles from the baselines from which the breadth of the territorial sea is measured (Art. 57 Part 5 UNCLOS). There are certain rights, jurisdiction and duties of the coastal State in the exclusive economic zone such as specific sovereign rights and the duty to regard the rights of other States (Art. 56 Part 5 UNCLOS). Though the UNCLOS has 168 member parties, the U.S. still not ratified the law because of its disagreement about Part 6. of the convention, which deals with the aspect of minerals found on the seabed in the EEZ and its distribution.[21] Nonetheless the USA is contracting party of the previous contract from 1958. Moreover, many parts of the UNCLOS are customary international law, meaning they emerge through general practice, based on the conviction that the norm is legally binding.[22] In part 7. the UNCLOS contains a own passage that deals with the protection and preservation of the marine environment, binding all member states and oblige them to prevent pollution. States have the obligation to protect and preserve the marine environment (Art. 192 Part 7 UNCLOS), States shall take, […] all measures consistent with the Convention that are necessary to prevent, reduce and control pollution […] (Art 194 Part 7. UNCLOS). This pollution may come from seabed activities (Art. 208), activities in the area (Art. 209), from dumping (Art. 210), from vessels (Art. 211) or from the atmosphere (Art. 212). Another International Convention to protect the environment from pollution through ships is the International Convention for the Prevention of Marine Pollution from Ships (MARPOL) from 1973. This Convention is completed by the International Convention on Civil Liability for Oil Pollution Damage which was revised 1992.[23]

---

[19] Von Arnauld, Andreas „Völkerrecht", 4. Auflage, 2019, p. 364.
[20] https://www.un.org/depts/los/convention_agreements/convention_historical_perspective.htm#Historical%20Pe rspective
[21] https://www.marineinsight.com/maritime-law/nautical-law-what-is-unclos/
[22] Von Arnauld, Andreas „Völkerrecht", 4. Auflage, 2019, p. 364.
[23] Krajewski, Markus „Völkerrecht", 2. Auflage, 2020, p. 370.

Besides these International Laws, there are federal laws of the USA to protect the environment like the Oil Pollution Act from 1990, issued after the Exxon Valdez accident and the Clean Water Act from 1977, as a result of the growing public awareness and concern for controlling water pollution, it was governing pollution control and water quality of the Nation's waterways.[24] The dilemma of ocean governance is that the oceans are beyond the remit of any single government to protect and there are different actors involved with different interests. As this paper approaches the consequences of an oil spill in the context of an oil exploration, it is necessary to identify who is involved and who is responsible for the offshore oil industry. This will be shown in the next section of this chapter.

The offshore oil industry is a vast configuration of structures, vessels and companies. The responsibility can be split in four phases of activity: exploration, development (drilling), production and decommissioning. First petroleum has to be identified and a lease has to be given from the federal government. The developer drills exploratory wells trying to locate oil or gas. If the search is successful, the well is temporarily sealed. Further preparations are made in order to extract the petroleum from the sea floor and a method is developed to transport it back to the shore.[25] Several federal government agencies like the U.S. Department of the Interior's Bureau of Ocean Energy Management and the Bureau of Safety and Environmental Enforcement, organize the natural resources in the EEZ. The management of offshore energy resources is done by these agencies while private companies possess the functional responsibility. The companies pay royalties to the government on the energy resources they produce from the leased areas. It is mainly oil and natural gas that is drilled on the ocean floor.[26] The list of companies operating in the Gulf of Mexico is long, among them the largest and most influential oil companies of the world: ExxonMobile, the largest non-governement owned integrated oil company and the largest oil company of the world producing 3% of the world's oil, Shell the second largest energy companies with employees in more than 100 countries and territories and BP the United Kingdom's largest corporation and leading operator in the Deepwater Gulf of Mexico.[27]

---

[24] https://www.boem.gov/environment/environmental-assessment/clean-water-act-cwa
[25] Bureau Ocean Energy Management „The Offshore Petroleum Industry in the Gulf of Mexico: A Continuum of Activities" p. 1.
[26] https://www.eia.gov/energyexplained/oil-and-petroleum-products/offshore-oil-and-gas.php
[27] https://www.offshoreinjuryfirm.com/offshore-injuries/company-profiles/

2.3 The Incident

Deepwater Horizon was a semi-submersible Mobile Offshore Drilling Unit that was owned by Transocean, the world's largest contractor of offshore drilling rigs. BP leased the Drilling Unit for 1 million US Dollars per day. Deepwater Horizon floated in 4,992 feet of water beyond the gentle slope of the continental shelf in the Mississippi Canyon to drill the Macondo well, a well that had proved complicated and challenging. On April 20th, BP and the Macondo well were almost six weeks behind schedule and over budget because Hurricane Ida battered the former rig that was supposed to drill the well. That day around 126 people were aboard the majority of them Transocean employees, a few BP men, cafeteria and laundry workers and workers for specialized jobs depending on the status of the well. The oil field, that the rig was drilling was located two and a half miles below the seabed in the Middle Miocene era trapped in a porous rock formation at temperatures exceeding 200 degrees.[28] The field was in the ultra-deepwater and drilling it was more challenging (see above). The Deepwater Horizon was registered at the Marshall Islands.[29]

BP leased the Mississippi Canyon Block 252, were the Macondo well is located, in the Central Gulf of Mexico at Minerals Management Service (MMS). The MMS was an agency of the United States Department of the Interior that managed the nations gas, oil and mineral resources on the outer continental shelf. After the Deepwater Horizon incident, the agency was split in the Bureau of Ocean Energy Management and the Bureau of Safety and Environmental Enforcement (named above). The 10 years lease started 2008. BP shared its ownership with Anadarko Petroleum and MOEX Offshore. The Macondo well is approximately 48 miles from the nearest shoreline, 114 mils from the shipping supply point of Port Fourchon, Louisiana. The well was an infrastructure-led development, it was designed so that it could later be a production well. By the time the accident happened the rig crew was preparing for the final activities associated with temporary well abandonment.[30] The final string of casing hat been run into the well, the cement barrier hat been put in place to isolate the hydrocarbon zones. Furthermore, integrity test had been conducted and the top of mud was being circulated out. Now, as final steps the cement plug had to be set in the casing and the lockdown sleeve had to be installed on the casing hanger seal assembly prior to disconnect the BOP and suspend the well.[31]

---

[28] National Commission on the BP Deepwater Horizon Oil Spill and Offshore Drilling „Deepwater – The Gulf Oil Disaster and the Future of Offshore Drilling", 2011, p. 2 f.

[29] Kirchner, Stefan and Alkanli, Deniz „Staatenverantwortlichkeit und völkerrechtlicher Meeresumweltschutz: Deepwater Horizon", 2011, p. 29.

[30] BP „Deepwater Horizon Accident Investigation Report", 2011, p. 15 f.

[31] BP „Deepwater Horizon Accident Investigation Report", 2011, p. 18.

The well had been drilled up to 18,360 ft. and sealed with cement to further protect against the chance of a leak or blowout. The cement mixture used for this well was designed, tested and provided by the firm Halliburton.[32] The "negative pressure" test was made and although there were discussions on the interpretation of the results the experts agreed that the Macondo was stable. However unexpected oil and gas was seeping into the well while the crew was pumping mud from it to reduce the weight bearing down on the hydrocarbons. It was racing up the Horizon's riser pipe pushing the drilling mud up from the well, just like a water fountain. Mud and water exploded up inside the derrick. Gas alarms began lighting up as gas spread over the rig. The blowout preventer was used but was incapable of preventing this blowout. The gas inflames causing explosion and fire.[33] In the report revealing the failures that lead to the blow out BP presents eight key findings:

1. The annulus cement barrier did not isolate the hydrocarbons.
2. The shoe track barriers did not isolate the hydrocarbons.
3. The negative pressure test was accepted although well integrity had not been established.
4. Influx was not recognized until hydrocarbons were in the riser.
5. Well control response actions failed to regain control of the well.
6. Diversion to the mud gas separator resulted in gas venting onto the rig.
7. The fire and gas system did not prevent hydrocarbon ignition.
8. The BOP emergency mode did not seal the well.[34]

Without these failures a blowout could have been prevented. Understanding who is responsible for the accident is not just identifying who is the owner of the platform, although even that is not easy to answer, as this paper has shown that the rig was leased. Further the crew working on it was from different companies and the permission to drill was giving by a state agency. After showing the economical and environmental consequences of the spill, this paper will identify who can be held responsible and how the spill changed policy.

2.4 The Consequences

11 people died that day, 17 were injured and 4 million barrels of crude oil spilled into the Gulf of Mexico causing a significant environmental impact, some of which is still being felt today.

---

[32] https://www.pmi.org/learning/library/comparison-risk-events-with-risk-management-9919
[33] https://www.nytimes.com/2010/12/26/us/26spill.html
[34] BP „Deepwater Horizon Accident Investigation Report", 2011, p. 31 f.

Experts estimate that around 205.8 million gallons (4.9 million barrels) of oil entered the Gulf of Mexico. That is about 19 times the size of the Exxon Valdez accident (see above).[35] The spill left an oil slick on the surface of the ocean 100 miles wide. The oil effected marine life, birds and groundwater. In the area a large number of fish kills were reported, and they had more lesions and sores than average.[36] Fish and many invertebrates absorb petroleum hydrocarbons through the skin and the gills. As a result of oil contamination, the plumage of birds can no longer perform its vital functions of repelling water and thermal insulation. The same effect occurs on the fur of marine mammals that is coated with oil. They ingest the oil when they attempt to clean their oil soiled feathers or fur. Oil disaster impair the growth of mussels and fish and a behavioural change can be observed. Poisoning by oil can cause genetic damage, increase ovum mortality, malformation, the number of genetic mutation and can damage reproductive organs. Moreover, many marine faunae lose their sense of direction, as many living organisms use very fine concentrations of substances to find their way around their environment and the alien hydrocarbons that enter the water may distract the animal from detecting the natural substances. As a consequence, they have difficulties finding food or identifying a breeding partner.[37] The regeneration periods depend on the character of the shore, protected soft substrates, salt marshes and mangroves have for example more difficulties to regenerate from oil spills than exposed rocky and sandy shores. The regeneration periods rank from between a few months up to more than 20 years.[38] Countries respond differently to oil spills (from the technical perspective), while some use exclusively mechanical methods such as oil skimmers or booms that form floating barriers, other opt for chemical methods such as dispersants. After the Deepwater Horizon blowout cleaning up measures were taken including booms and skimming off oil, collecting tar balls manual and wiping of marsh grass. Dispersants were also used in large amounts.[39] However, their effectiveness heavily depend on the type and condition of the oil. Following the explosion, the oil flowed out of the borehole in great depth and entered the water as a massive cloud. So far, little experience has been gained in responding to accidents on this scale.[40] Fishermen involved in the cleaning effort often criticize that dispersants are more damaging to the ecosystem than the oil itself, as oil is a natural product and dispersants aren't.

[35] Krupnick, Alan; Campell, Sarah; Cohen, Mark A. and Parry, Ian W.H. „Understanding the Costs and Benefits of Deepwater Oil Drilling Regulations", 2011, p. 4.
[36] https://www.natlenvtrainers.com/blog/article/the-environmental-impact-of-the-deepwater-horizon-oil-spill
[37] World Ocean Review "Last stop: The ocean – polluting the seas", 2010, p. 96.
[38] World Ocean Review „Last stop: The ocean – polluting the seas", 2010, p. 97.
[39] Harzl, Viktoria and Pickl, Matthias „The Future of Offshore Oil Drilling – An Evaluation of the Economic, Environmental and Political Consequences of the Deppwater Horizon Incident", 2012, p. 761.
[40] World Ocean Review „Last stop: The ocean – polluting the seas", 2010, p. 97.

BP deployed many vessels, barges and skimmers many of them owned and operated by locals. Some locals hired for the clean-up by BP received up to 2000 Dollar per day for their boats. Cleaning for BP was therefore for them more profitable than fishing or tourism. Fishers who had contracts with BP for cleaning could take advantage and were able to earn significantly more than in a regular year of fishing however, for those who had no contracts the oil spill came as a double blow as they already suffered losses after Hurricane Katrina. Prices of shrimp, fish and oysters from the affected regions gone up since catches have decreased. The loss was not only attributed to decreasing fishing quota but also to people's perception of product safety. The disaster increased traffic from additional clean-up crews, press and officials which was another kind of tourism which had a positive effect on the hotel industry. Although the effect was short-term only, the overall booking decreased and hotels and restaurants experienced cancellations. Some people have benefited from the spill, while most gulf residents suffered and will suffer financial hardship for years to come.[41] The compensation of losses generally depends on the level of documentation and proof the injured parties have to bring and in fact moreover legal standards often make it difficult for victims to prove damages what leaves many of them undercompensated. On the other hand, it is also possible that some individuals were overcompensated. Furthermore, from an economic welfare perspective only direct harms count as welfare losses. In conclusion lost income in the fishing and tourism industry is not considered to be an economic damage and thus is not needed to be compensated from the firm responsible for the oil spill.[42]

3. Political and Legal Response

3.1 Responsibility

If an oil spill occur the firm is legally obligated under the Oil Pollution Act (see above) to try to contain further spillage as well as clean-up. However, there is no dollar cap on liability for clean-up. Measures might include booms, spraying dispersants, mopping up beaches and rehabilitating affected wildlife. If a firm is unable to pay the cost, funds from the Oil Pollution Liability Trust Fund are tapped.[43]

---

[41] Harzl, Viktoria and Pickl, Mathhias „The Future of Offshore Oil Drilling – An Evaluation of the Economic, Environmental and Political Consequences of the Deppwater Horizon Incident", 2012, p. 762 ff.
[42] Krupnick, Alan; Campell, Sarah; Cohen, Mark A. and Parry, Ian W.H. „Understanding the Costs and Benefits of Deepwater Oil Drilling Regulations", 2011, p. 42.
[43] Krupnick, Alan; Campell, Sarah; Cohen, Mark A. and Parry, Ian W.H. „Understanding the Costs and Benefits of Deepwater Oil Drilling Regulations", 2011, p. 41.

As BP stated in the investigation report "The team did not identify any single action or inaction that caused this accident. Rather, a complex and interlinked series of mechanical failures, human judgments [...] came together to allow [...] the accident. Multiple companies, work teams and circumstances were involved over time."[44] BP is ascribing responsibility to Transocean, the firm that owns the ring and was responsible for the blowout preventer that failed. Further, Halliburton the firm that provided the cement is accused as the cement prove to be weak designed and not enough tested for quality and risk assessment. Finally, the crew who were not only BP workers but also Transocean staff and specialists working for Halliburton contributed or fall short in preventing the blow out.[45]

BP spend years of litigation reaching an agreement with the Justice Department Manslaughter, plead guilty to 14 criminal counts including Environmental Crimes, Obstruction of Congress and violation of the Clean Water Act. The record civil settlement with BP and the Justice Department and Gulf coast states amounts over 20 billion Dollar, included billions of Dollars penalty under the Clean Water, in natural resource damages and payments to states (Alabama, Florida, Louisiana, Mississippi and Texas) and local governments. Further BP paid companies and individuals harmed by the spill.[46] Transocean plead guilty to Environmental Crime and had to pay 1 billion in civil penalties and 400 million in criminal fines what appropriately reflects its role in the disaster.[47] Moreover, Transocean and Halliburton agreed on settlements for individuals, business and local government seeking compensation over the spill.[48] BP spokesman Geoff Morell is referring to the settlement by Halliburton saying that finally Halliburton has acknowledged that it played a role in the accident.[49] However, BP is the company who is mainly responsible of the accident. U.S. District Judge Carl Barbier attributed 67% of fault to BP, 30% to Transocean and 3% to Halliburton. He found BP's conduct as reckless and Transocean's and Halliburton's conduct as negligent.[50] The Project Management Institute classify BP as risk seeking organization because of a deeply-entrenched culture of valuing cost over quality, minimal financial repercussions in operational failure, minimal independent oversight of operations, and deep cash reserves. The driving force behind BP's decisions and risk tolerance were revenues and cost cutting. Quality seems to be the one constraint that could be sacrificed. Although quality in this case means quality assurance.

---

[44] BP „Deepwater Horizon Accident Investigation Report", 2011, p. 11.
[45] https://www.bbc.com/news/world-us-canada-11230757
[46] https://theconversation.com/bp-paid-a-steep-price-for-the-gulf-oil-spill-but-for-the-us-a-decade-later-its-business-as-usual-136905
[47] https://www.noaa.gov/explainers/deepwater-horizon-oil-spill-settlements-where-money-went
[48] https://www.ft.com/content/74842ae0-ff89-11e4-8c46-00144feabdc0
[49] https://www.dw.com/en/halliburton-to-pay-victims-of-deepwater-horizon-oil-spill/a-17896503
[50] https://time.com/3268978/bp-gulf-oil-spill-louisiana-court/

BP had several proactive measures they could have invested in to minimize a potential blowout like a relief well, an Acoustic Switch (an additional means of actuating the BOP) or a Capping Stock (a backup device in case the BOP fail), yet these measures were costly and time consuming. BP's cash position was great and in the past years, they were rebounding from disaster nicely, so the company saw little cause spending millions in preventative measures.[51] However, it is not just the companies related to the oil drill that can be seen as responsible for what happened. As the author worked out in the previous chapters the government itself leased the drilling spot to BP. The Minerals Management Service, the agency that should oversee the sea drilling operations in the Gulf, was understaffed and under-equipped to perform an efficient monitoring.[52] Further it is the government that is held responsible under the UNCLOS for its exclusive economic zone and activities going on in its territory. In contrast to tanker disasters, which often occur near the coast, the Deepwater Horizon oil spill is not a locally limited event. It affects large parts of the Atlantic coast, US states and the greater area of the Gulf of Mexico. Even if the majority of the damage occurred on the US south coast, the question of liability towards other states such as Cuba remains. The cause of damage (the leaking borehole) lay in the EEZ of the U.S. and is therefore subject to the regulation authority of the United States. However, the Marshall Islands were allowed to regulate the technical security theoretical, there weren't able to guide the concrete mission of the platform. The drilling in the Gulf of Mexico took place in the EEZ of the USA and therefore the USA had the authority to observe and regulate.[53] The United States were allowed to exploit crude oil deposits (Art. 56 UNCLOS) but obliged to prevent pollution (Art. 208 UCLOS) (see above). To avoid environmental damage, especially those that could affect others results from the generally permissible use of the EEZ. Even if the UNCLOS is not ratified by the USA that duty can be drawn as well from the principle of good neighborhood put down in Art. 74 of the UN-Charta "Members of the United Nations also agree that their policy in respect of the territories [...] must be based on the general principle of good-neighborliness [...]".[54] Consequently the US can be accused of violating international law. In this case the damage is not caused by the state but rather by non-state companies, nevertheless the state authorized the drilling in the Macondo well and had the factual control. They have failed to effectively protect the marine environment in the EEZ.[55]

---

[51] https://www.pmi.org/learning/library/comparison-risk-events-with-risk-management-9919
[52] https://www.pmi.org/learning/library/comparison-risk-events-with-risk-management-9919
[53] Kirchner, Stefan and Alkanli, Deniz „Staatenverantwortlichkeit und völkerrechtlicher Meeresumweltschutz: Deepwater Horizon", 2011, p. 28 ff.
[54] See Ibid., p. 35.
[55] See Ibid., p. 39.

3.2 Reaction Obama Administration

Authorities had often failed to adopt a prompt, effective or appropriate response to oil spill in the present due to a lack of clear responsibilities, equipment and personnel.[56]

The drilling risk was totally underestimated by politician, as Barack Obama justifies his position on offshore drilling only a few weeks before the Deepwater Horizon disaster by saying "[…] It turns out, by the way, that oil rigs today generally don't cause spills […].[57]

Broder who is writing for the New York Times is accusing the government of underestimating how much oil was flowing out into the Gulf of Mexico and how much left after the well was capped in July. Misinformation lead to a loss of faith in the government's ability to handle the spill and a continuing breach between the federal authorities and state and local officials. Further, the Obama administration failed to fully inform the public about its own worst-case estimates keeping the American people in the dark about the size of the disaster. For example, the government said that the flow rate was ca. 5,000 barrels of oil a day, even though BP officials and scientists acknowledged that the rate could be as high as 110,000 barrels a day.[58]

One month after the environmental disaster Barack Obama, who claimed before that BP alone was to blame for the catastrophe, was standing up to take command of the clean-up effort and told the nation that he takes responsibility. He acknowledged that his administration had failed to reform the Minerals Management Service, who became a scandal-ridden federal agency that allowed the oil industry to self-regulate. The president didn't see sufficient urgency to reform and underestimated the situation. The former Bush administration paved the way for the accident however Obama gave BP the green light to drill. Under Bush the MMS charged with safeguarding the environment from the ravage of drilling descended criminality. The staff of the agency was in close favourable of the oil industry they went to parties, trips and had private contact with oil-company officials. MMS managers received cash bonuses for pushing through risky offshore leases, they didn't investigate shady deals and even allowed the oil companies to fill in their won inspection reports.[59]

The government tried to support BP's cleaning efforts, deploying thousands of vessels to collect and contain the oil and almost two million gallons of dispersants.[60] If a government steps in to support the responsible firm, these expenses have to reimbursed.

---

[56] Ocean Review „Last stop: The ocean – polluting the seas", 2010, p. 98.
[57] https://shadowproof.com/2010/04/29/obama-oil-rigs-today-generally-don't-cause-spills/
[58] https://www.nytimes.com/2010/10/07/science/earth/07spill.html
[59] https://www.rollingstone.com/politics/politics-news/the-spill-the-scandal-and-the-president-193093/
[60] https://www.nytimes.com/2010/10/07/science/earth/07spill.html

On October 12, 2011 the U.S. Coast Guard's National Pollution Funds Centre is reported to have billed BP 581-million-dollar recovery costs related to Deepwater Horizon.[61]

The accident caused the review of the regulation structures that contributed to the missteps leading to the disaster. The Obama administration worked on balancing environmental and safety concerns with energy development including the reorganization of the MMS in order to improve their independence and strength of its inspection capabilities. Under Ken Salazar the Secretary of the Interior the MMS split into three separate bureaus. First an interim bureau, the Bureau of Energy Management, Regulation, and Enforcement (BOEMRE) replaced the MMS. BOEMRE put a series of new safety measures in place. Salazar announced the Drilling Safety Rule and Workplace Safety Rule that obtain new well bore integrity requirements, requirements related to the blowout preventer and control system.[62] Further the Safety and Environmental Management System 1 & 2 and the Blowout Preventer and Well Control Rule were published.[63] The Secretary of the Interior announced a moratorium on deepwater drilling which was lifted only if drillers could show compliance with all safety and environmental requirements. Later the BOEMRE split in the Bureau of Ocean Energy Management (BOEM) that became responsible for the evaluation, planning and leasing and the Bureau of Safety and Environmental Enforcement (BSEE), responsible for safety and environmental enforcement. With this reorganization conflicting priorities should be minimized.[64] The breakup of the MMS had a political effect yet the processes under which the now separate agencies functioned did not change in many material respects. Safety condition had improved under BSEE nevertheless a report by the Government Accountability Office in 2016 and 2017 showed that like the former agency BSEE is understaffed and characterised by poor leadership.[65] There were serious steps taken in order to make drilling safer however, some fall down as they never came to pass Congress.[66] As worked out above historical oil spills were followed by changing laws. Still, in this case no revolutionary law following the accident. In response to the Deepwater Horizon spill the Congress passed the RESTORE Act 2012. An act that served to ensure that civil penalties paid by BP to the federal government would be shared with gulf coast states.

---

[61] Krupnick, Alan; Campell, Sarah; Cohen, Mark A. and Parry, Ian W.H. „Understanding the Costs and Benefits of Deepwater Oil Drilling Regulations", 2011, p. 41.

[62] https://eelp.law.harvard.edu/2020/05/deepwater-horizon-ten-years-later-reviewing-agency-and-regulatory-reforms/

[63] James, Robert A. and Pulman, Stella „Deepwater Horizon: A Decade of Legal Impacts", 2020, p. 4.

[64] https://eelp.law.harvard.edu/2020/05/deepwater-horizon-ten-years-later-reviewing-agency-and-regulatory-reforms/

[65] James, Robert A. and Pulman, Stella „Deepwater Horizon: A Decade of Legal Impacts", 2020, p. 3 f.

[66] https://eelp.law.harvard.edu/2020/05/deepwater-horizon-ten-years-later-reviewing-agency-and-regulatory-reforms/

However, there was no new law made on drilling safety or increasing energy companies' liability limits for oil spills.[67] The spill helped to create new agencies and new regulations and reinvigorated the discussion about America's addiction to oil and spurred the urge for investment in green technologies. It active environmentalists and locals to keep an eye on all the events that happen in the area even if the oil spill does not make daily headlines any more.[68] The scientists are increasingly encouraging world leaders to consider that any new oil development is unwise because emissions from fossil fuels exacerbate global heating. While Obama raised safety rules, the Trump administration has weakened those standards again encouraging the oil industry to take risks.[69]

4. Trump and the Future of Offshore drilling

Other than Obama, who experienced the accident as the president in office and had to deal with the aftermath, Trump and his position and policy regarding the oil industry was not that much under focus. However, other than the ex-president and his administration the Trump administration is known for encourage offshore oil and gas development and expanded offshore drilling.[70] Hence he is stepping in the shoes of conservative Republican that have backed own oil and gas production.[71] Trump is following a pro-job, pro-national security, and energy independence agenda.[72] He changed the Well Control Rule in 2019 in order to reduce the burden it comprises for the industry. The revised rule made changes in the regulations relating to reporting obligation, drilling margins, monitoring and more and therefore encourage energy exploration and production by reducing regulatory obstacles.[73] In a statement from September 2019 he is claiming, that the US does not need Middle Eastern oil anymore as the United States itself become number one energy producer in the world. But the statement is only half true as the United States experienced a technology driven drilling boom, but crude oil and petroleum products still flowed in abundantly. The U.S. is importing from Saudi Arabia, Iraq and other gulf nations because several refineries prefer their oil.

[67] https://theconversation.com/bp-paid-a-steep-price-for-the-gulf-oil-spill-but-for-the-us-a-decade-later-its-business-as-usual-136905
[68] Harzl, Viktoria and Pickl, Mathhias „The Future of Offshore Oil Drilling – An Evaluation of the Economic, Environmental and Political Consequences of the Deppwater Horizon Incident", 2012, p. 768.
[69] https://www.theguardian.com/environment/2020/apr/20/deepwater-horizon-10-years-later-could-it-happen-again
[70] https://www.voanews.com/usa/us-politics/trump-bans-oil-drilling-florida-georgia-south-carolina
[71] https://theconversation.com/trump-greenlights-drilling-in-the-arctic-national-wildlife-refuge-but-will-oil-companies-show-up-144715
[72] https://www.whitehouse.gov/briefings-statements/president-donald-j-trump-unleashed-american-producers-restored-energy-dominance/
[73] James, Robert A. and Pulman, Stella „Deepwater Horizon: A Decade of Legal Impacts", 2020, p. 4.

The mismatch between what the U.S. refiners want and what the states produces means that million barrels of crude oil and petroleum have to be imported every month.[74] In August 2020 Trump announced the opening of the Arctic National Wildlife Refuge to oil and gas development. This action may come from his desire to be independent from "foreign oil".[75] However Trumps policy seem to be unpredictable as he recently extends the drilling ban in the coast of Florida, Georgia and South Caroline, a great success for environmentalists as the current moratorium on drilling in that areas would have expired in 2022. The industry officials were surprised by Trumps action as nobody seemed to know where the idea of extending the moratorium came from. In 2017 Trump still was in favour of the oil industry and announced that his administration would seek to open nearly all U.S. coastal waters to oil and gas drilling. But he changed his mind maybe to push his campaign, as blocking drilling is good electoral politics.[76] Gina McCarthy the president of the Natural Resource Defense Council Action Fund is convinced that Trump used the extension of the moratorium in order to win votes she describes his act as "[...] a transparent attempt to manipulate Floridians two month before Election Day".[77]

Today the oil industry is facing great challenges: a collapse in oil demand due to the global pandemic, new uncertainty about the future of global oil demand because of alternatives such as electric vehicles and new environmental laws, the possibility of Democratic victory in the November 2020 election and the likeliness of policy that aim to reduce fossil fuel use and increasing investors pressure on banks and investment firms to reduce support for fossil fuel projects.[78] There is a greater ecological sensibility in 2020 that it was ten years ago when the Deepwater Horizon accident happened. And big change in safety culture has been achieved. However, experts are sure that another blow is very likely to happen again as drilling operations are never completely safe. The risk of a blowout can be reduced but as long as oil is drilled and transported in the oceans, there will be the danger that an accident happens.[79] The next presidency may decide over the future of the oil industry.

---

[74] https://www.reuters.com/article/us-saudi-aramco-attacks-trump/trump-says-u-s-does-not-need-middle-east-oil-but-cargoes-keep-coming-idUSKBN1W12RO
[75] https://theconversation.com/trump-greenlights-drilling-in-the-arctic-national-wildlife-refuge-but-will-oil-companies-show-up-144715
[76] https://www.politico.com/news/2020/09/08/trump-oil-drilling-florida-410
[77] https://www.voanews.com/usa/us-politics/trump-bans-oil-drilling-florida-georgia-south-carolina
[78] https://theconversation.com/trump-greenlights-drilling-in-the-arctic-national-wildlife-refuge-but-will-oil-companies-show-up-144715
[79] https://www.theguardian.com/environment/2020/apr/20/deepwater-horizon-10-years-later-could-it-happen-again

Candidate Biden has pledged to take action against the energy companies just like the Democratic presidents before him that promoted an anti-fossil fuel policy.[80] Trump seems to be the better option for the oil industry however this paper showed that his actions were contradictory. His unreliable character makes him unpredictable nevertheless his anger of import oil and aim for Americas independence gives grounds for optimism in the offshore industry.

5. Conclusion

This paper has shown the development of the oil industry in America and the international attempts for ocean governance that prevent pollution from oil. The international customary law written down in the UNCLOS ascribes accountability to countries for them using their exclusive economic zones. The U.S. further enact laws that regulate offshore oil drilling like the Oil Pollution Act from 1990 that regulates the dealing with an oil-accident and input liability to firms involved. However, safety lacks, corruption, human failures and the obsessive search of profit lead to the accident of the Deepwater Horizon, the biggest oil spill in history. The environmental and economical-consequences were great reducing fish resources and leaving many hotels on the affected beaches bankrupt. BP was mainly ascribed responsibly for the spill nevertheless other companies were involved as well and had to pay for their failures, like Halliburton the firm that provided the cement and Transocean the owner of the Deepwater Horizon platform. Moreover, the government (more precisely MMS) played its part in the accident neglecting its task to monitor the drilling. Obama saw the failures made under his policy and the policy of the pre-administration. He reformed the rotten structures promoting safety and damming offshore drilling. Despite no revolutionary laws were made the accident changed the perception of the oil industry and sensitized society and politics. What the incident did was bring attention to the mismanagement in regulating the companies. The accident resulted in an up-great of safety rules and a growing number of environmentalists. As this paper has shown Trumps policy mainly confronting Obamas achievements and encouraging the oil industry to expand drilling. The new president is confronted with a growing anti-oil-lobby as well. The next presidential election will determine the future of the offshore industry. However, this future does not seem to be too rosy for the petroleum sector, in the context of growing environmentalism and the development of green technologies that shall replace oil.

---

[80] https://www.eenews.net/stories/1061279307

Bibliography

Bacevich, Andrew J. „The limits of power", 2008, New York: Metropolitan Books.

BP „Deepwater Horizon Accident Investigation Report", September 8, 2011.

Bureau Ocean Energy Management „The Offshore Petroleum Industry in the Gulf of Mexico: A Continuum of Activities".

Harzl, Viktoria and Pickl, Matthias „The Future of Offshore Oil Drilling – An Evaluation of Economic, Environmental and Political Consequences oft he Deewater Horizon Incident" in: *Energy & Environment*, Vol. 23, No. 5, 2012, Essex: Multi-Science   Publishing co. ltd.

James, Rober A. and Pulman, Stella „Deepwater Horizon: A Decade of Legal Impacts", pillsburylaw, 2020.

Kirchner, Stefan and Alkanli, Deniz „Staatenverantwortlichkeit und völkerrechtlicher Meeresumweltschutz: Deepwater Horizon", 2011, StuZR.

Krajewski, Markus „Völkerrecht", 2. Auflage, 2020.

Krupnick, Alan; Campell, Sarah; Cohen, Mark A. and Parry, Ian W.H. „Understanding the Costs and Benefits of Deepwater Oil Drilling Regulations" in: *Resources for the Future*, January 2011, Washinton DC.

National Commission on the BP Deepwater Horizon Oil Spill and Offshore Drilling – Report to the President „Deepwater – The Gulf Oil Disaster and the Future of Offshore Drilling",January 2011.

Singh, Pradeep A. and Ort, Mara „Law and Policy Dimensions of Ocean Governance" in: Jungblat S. et al. (eds.) *YOUMARES 9 – The Oceans: Our Research, Our Future*,2020, Springer.

Von Arnauld, Andreas „Völkerrecht", 4. Auflage, 2019.

World Ocean Review „Last stop: The ocean – polluting the seas", 2010.

Arnold & Itkin LLP Trial Lawyers „The Leaders in Martitime Law", online available: https://www.offshoreinjuryfirm.com/offshore-injuries/company-profiles/ (28.09.20).

BBC „Who's blamed by BP fort he Deepwater Horizon spill?", 2010, online available: https://www.bbc.com/news/world-us-canada-11230757 (28.09.20).

BOEM „Clean Water Act (CWA)", online available: https://www.boem.gov/environment/environmental-assessment/clean-water-act-cwa (28.09.20).

BOEM „The Oil Pollution Act of 1990", online available: https://www.boem.gov/sites/default/files/documents//The%20Oil%20Pollution%20Act%20of%20199 0.pdf (28.09.20).

DW „Halliburton to pay victims of Deepwater Horizon oil spill", 2014, online available: https://www.dw.com/en/halliburton-to-pay-victims-of-deepwater-horizon-oil-spill/a-17896503 (28.09.20).

EIA „Oil and petroleum products explained – Offhore oil and gas", 2019, online available: https://www.eia.gov/energyexplained/oil-and-petroleum-products/offshore-oil-and-gas.php (28.09.20).

EIA„Gulf of Mexico Fact Sheet", online available: https://www.eia.gov/special/gulf_of_mexico/ (28.09.20).

Environmental & Energy Law Program „Deepwater Horizon Ten Years Later: Reviewing agency regulatory reforms" by Vizcarra, Hana, 2020, online available: https://eelp.law.harvard.edu/2020/05/deepwater-horizon-ten-years-later-reviewing-agency-and-regulatory-reforms/ (28.09.20).

E&E News „Oil faces existential risk from Democratic president. Right?" by Metthews, Mark K. and Cama, Timothy, 2019, online available: https://www.eenews.net/stories/1061279307 (28.09.20).

Financial Times „BP settles with Transocean and Halliburton over gulf spill" by Crooks, Ed, 2015, online available: https://www.ft.com/content/74842ae0-ff89-11e4-8c46-00144feabdc0 (28.09.20).

Fortune „A short history of drilling in the Gulf of Mexico" by Burke, Doris, 2011, online available: https://fortune.com/2011/01/24/a-short-history-of-drilling-in-the-gulf-of-mexico/ (28.09.20).

History „Exxon Valdez Oil Spill", 2018, online available: https://www.history.com/topics/1980s/exxon-valdez-oil-spill (28.09.20).

Los Angeles Times „The 1969 Santa Barbara oil spill that changed oil and gas exploration forever" by Mai-Duc, Christine, 2015, online available: https://www.latimes.com/local/lanow/la-me-ln-santa-barbara-oil-spill-1969-20150520-htmlstory.html (28.09.20).

Marine insight „Natuical Law: What is UNCLOS?" by sharda, 2019, online available: https://www.marineinsight.com/maritime-law/nautical-law-what-is-unclos/ (28.09.20).

National Environmental Trainers inc. „The Environmental Impact oft he Deepwater Horizon Oil Spill", online available: https://www.natlenvtrainers.com/blog/article/the-environmental-impact-of-the-deepwater-horizon-oil-spill (28.09.20).

National Oceanic and Atmospheric Administration „Explosion triggered economic, environmental devastation, and a legal battle", 2017, online available: https://www.noaa.gov/explainers/deepwater-horizon-oil-spill-settlements-where-money-went (28.09.20).

Ocean & Law of the Sea – United Nations „The United Nations Convention on the Law of the Sea (A historical perspective)" online available: https://www.un.org/depts/los/convention_agreements/convention_historical_perspective.htm#Historical%20Perspective (28.09.20).

Politico „Trump expands oil drilling moratorium for Florida" by Lefebvre, Ben and Colman, Zack, 2020, online available: https://www.politico.com/news/2020/09/08/trump-oil-drilling-florida-410042 https://www.voanews.com/usa/us-politics/trump-bans-oil-drilling-florida-georgia-south-carolina (28.09.20).

Project Management Institute „Deppwater Horizons lessons in probabilities – The Project Solvers of America, Inc." by Greene-Blose, Johanne M., 2015, online available: https://www.pmi.org/learning/library/comparison-risk-events-with-risk-management-9919 (28.09.20).

Reuters „Trump says U.S. does not need Middle East oil, but cargoes keep coming" by Gardner, Timothy, 2019, online available: https://www.reuters.com/article/us-saudi-aramco-attacks-trump/trump-says-u-s-does-not-need-middle-east-oil-but-cargoes-keep-coming-idUSKBN1W12RO (28.09.20).

Rolling Stone „The Spill, The Scandal and the President" by Dickinson, Tim, 2010, online available: https://www.rollingstone.com/politics/politics-news/the-spill-the-scandal-and-the-president-193093/ (28.09.20).

Shadow Proof „Obama: Oil Rigs today generally don't cause spills", 2010, online available: https://shadowproof.com/2010/04/29/obama-oil-rigs-today-generally-don't-cause-spills/ (28.09.20).

The Conversation „Trump greenlights drilling in the Arctic National Wildlife Refuge, but will oil companies show up?", 2020, online available: https://theconversation.com/trump-greenlights-drilling-in-the-arctic-national-wildlife-refuge-but-will-oil-companies-show-up-144715 (28.09.20).

The Conversation „BP paid a steep price fort he Gulf oil spill but fort he US a decade later, it's business as usual" by Uhlmann, David M., 2020, online available: https://theconversation.com/bp-paid-a-steep-price-for-the-gulf-oil-spill-but-for-the-us-a-decade-later-its-business-as-usual-136905 (28.09.20).

The Guardian „Of course it could happen again: experts say little has changed since Deepwater Horizon" by Holden, Emily, 2020, online available: https://www.theguardian.com/environment/2020/apr/20/deepwater-horizon-10-years-later-could-it-happen-again (28.09.20).

The New York Times „Report Slams Administration for Undererstimating Gulf Spill" by Broder, John M., 2010, online available: https://www.nytimes.com/2010/10/07/science/earth/07spill.html (28.09.20).

The New York Times „Deepwater Horizon's Final Hours" by Barstow, David; Rohde, David and Saul, Stephanie, 2010, online available: https://www.nytimes.com/2010/12/26/us/26spill.html (28.09.20).

Tiger General „Advantages of Offshore Oil Rigs and Drilling" by Overholt, Mark, 2017, online available: https://www.tigergeneral.com/advantages-offshore-oil-rigs-drilling/ (28.09.20).

Time „Judge Places Most Blame on BP for 2010 Oil Spill" by Rayman, Noah, 2014, online available: https://time.com/3268978/bp-gulf-oil-spill-louisiana-court/ (28.09.20).

VOA „Trump Bans Oil Drilling Off Florida, Georgia, South Garolina" by Baragona, Steve, 2020, online available: https://www.voanews.com/usa/us-politics/trump-bans-oil-drilling-florida-georgia-south-carolina (28.09.20).

White House „President Donald J. Trump Has Unleashed American Producers and Restored Our Energy Dominance", 2020, online available: https://www.whitehouse.gov/briefings-statements/president-donald-j-trump-unleashed-american-producers-restored-energy-dominance/ (28.09.20).